Use of the Dead to the Living

Southwood Smith

Alpha Editions

This edition published in 2024

ISBN : 9789362093790

Design and Setting By
Alpha Editions
www.alphaedis.com
Email - info@alphaedis.com

As per information held with us this book is in Public Domain.
This book is a reproduction of an important historical work. Alpha Editions uses the best technology to reproduce historical work in the same manner it was first published to preserve its original nature. Any marks or number seen are left intentionally to preserve its true form.

USE OF THE DEAD TO THE LIVING.

FROM THE WESTMINSTER REVIEW.

An Appeal to the Public and to the Legislature, on the necessity of affording Dead Bodies to the Schools of Anatomy, by Legislative Enactment. By WILLIAM MACKENZIE. Glasgow. 1824.

Every one desires to live as long as he can. Every one values health "above all gold and treasure." Every one knows that as far as his own individual good is concerned, protracted life and a frame of body sound and strong, free from the thousand pains that flesh is heir to, are unspeakably more important than all other objects, because life and health must be secured before any possible result of any possible circumstance can be of consequence to him. In the improvement of the art which has for its object the preservation of health and life, every individual is, therefore, deeply interested. An enlightened physician and a skilful surgeon, are in the daily habit of administering to their fellow men more real and unquestionable good, than is communicated, or communicable by any other class of human beings to another. Ignorant physicians and surgeons are the most deadly enemies of the community: the plague itself is not so destructive; its ravages are at distant intervals, and are accompanied with open and alarming notice of its purpose and power; theirs are constant, silent, secret; and it is while they are looked up to as saviours, with the confidence of hope, that they give speed to the progress of disease and certainty to the stroke of death.

It is deeply to be lamented that the community, in general, are so entirely ignorant of all that relates to the art and the science of medicine. An explanation of the functions of the animal economy; of their most common and important deviations from the healthy state; of the remedies best adapted to restore them to a sound condition, and of the mode in which they operate, as far as that is known, ought to form a part of every course of liberal education. The profound ignorance of the people on all these subjects, is attended with many disadvantages to [Pg 4]themselves, and operates unfavorably on the medical character. In consequence of this want of information, persons neither know what are the attainments of the man in whose hands they place their life, nor what they ought to be; they can neither form an opinion of the course of education which it is incumbent on him to follow, nor judge of the success with which he has availed himself of the means of knowledge which have been afforded him. There is one branch of medical education in particular, the foundation, in fact, on which the whole superstructure must be raised, the necessity of which is not commonly understood, but which requires only to be stated to be perceived. Perhaps it is impossible to name any one subject which it is of more importance that the community should understand. It is one in which every man's life is deeply implicated: it is one on which every man's ignorance or information will have a considerable influence. We shall, therefore, enter into it with some detail: we shall show the kind of knowledge which it is indispensable that the physician and surgeon should possess; we shall illustrate, by a reference to particular cases, the reason why this kind of knowledge cannot be dispensed with: and we shall explain, by a statement of facts, the nature and extent of the obstacles which at present oppose the acquisition of this knowledge. We repeat, there is no subject in which every reader can be so immediately and deeply interested, and we trust that he will give us his calm and unprejudiced attention.

The basis of all medical and surgical knowledge is anatomy. Not a single step can be made either in medicine or surgery, considered either as an art or a science without it. This should seem self evident, and to need neither proof nor illustration: nevertheless, as it is useful occasionally to contemplate the evidence of important truth, we shall show why it is, that there can be no rational medicine, and no safe surgery, without a thorough knowledge of anatomy.

Disease, which it is the object of these arts to prevent and to cure, is denoted by disordered function: disordered function cannot be understood without a knowledge of healthy function; healthy function cannot be understood without a knowledge of structure; structure cannot be understood unless it be examined.

The organs on which all the important functions of the human body depend, are concealed from the view. There is no possibility of ascertaining their situation and connections, much less their nature and operation, without inspecting the interior of this curious and complicated machine. The results of the mechanism are visible; the mechanism itself is concealed, and must be investigated to be perceived. The important operations of nature are seldom entirely [Pg 5]hidden from the human eye; still less are they obtruded upon it, but over the most curious and wonderful operations of the animal economy so thick a veil is drawn, that they never could have been perceived without the most patient and minute research. The circulation of the blood, for example, never could have been discovered without dissection. Notwithstanding the partial knowledge of anatomy which must have been acquired by the accidents to which the human body is exposed, by attention to wounded men, by the observance of bodies killed by violence; by the huntsman in using his prey; by the priest in immolating his victims; by the augur in pursuing his divinations; by the slaughter of animals; by the dissection of brutes; and even occasionally by the dissection of the human body, century after century passed away, without a suspicion having been excited of the real functions of the two great systems of vessels, arteries and veins. It was not until the beginning of the 17th century, when anatomy was ardently cultivated, and had made considerable progress, that the valves of the veins and of the heart were discovered, and subsequently that the great Harvey, the pupil of the anatomist who discovered the latter, by inspecting the structure of these valves; by contemplating their disposition; by reasoning upon their use, was led to suspect the course of the blood, and afterwards to demonstrate it. Several systems of vessels in which the most important functions of animal life are carried on—the absorbent system, for example, and even that portion of it which receives the food after it is digested, and which conveys it into the blood, are invisible to the naked eye, except under peculiar circumstances: whence it must be evident, not only that the interior of the human body must be laid open, in order that its organs may be seen; but that these organs must be minutely and patiently dissected, in order that their structure may be understood.

The most important diseases have their seat in the organs of the body; an accurate acquaintance with their situation is, therefore, absolutely necessary, in order to ascertain the seats of disease; but for the reasons already assigned, their situation cannot be learnt, without the study of anatomy. In several regions, organs the most different in structure and function are placed close to each other. In what is termed the epigastric region, for example, are situated the stomach, the liver, the gall bladder, the first portion of the small intestine, (the duodenum) and a portion of the large intestine (the colon); each of these organs is essentially different in structure and in use, and is liable to distinct diseases. Diseases the most diversified, therefore, requiring the most opposite treatment, may exist in the same region of the body; the discrimination of which is absolutely impossible, [Pg 6]without that knowledge which the study of anatomy alone can impart.

The seat of pain is often at a great distance from that of the affected organ. In disease of the liver, the pain is generally felt at the top of the right shoulder. The right phrenic nerve sends a branch to the liver: the third cervical nerve, from which the phrenic arises, distributes numerous branches to the neighborhood of the shoulder: thus is established a nervous communication between the shoulder and the liver. This is a fact which nothing but anatomy could teach, and affords the explanation of a symptom which nothing but anatomy could give. The knowledge of it would infallibly correct a mistake, into which a person who is ignorant of it, would be sure to fall: in fact, persons ignorant of it do constantly commit the error. We have know several instances in which organic disease of the liver has been considered, and treated as rheumatism of the shoulder. In each of these cases, disease in a most important organ might have been allowed to steal on insidiously, until it became incurable; while a person, acquainted with anatomy, would have detected it at once, and cured it without difficulty. Many cases have occurred of persons who have been supposed to labor under disease of the liver, and who have been treated accordingly: on examination after death, the liver has been found perfectly healthy, but there has been discovered extensive disease of the brain. Disease of the liver is often mistaken for disease of the lungs: on the other hand, the lungs have been found full of ulcers, when they were supposed to have been perfectly sound, and when every symptom was referred to disease of the liver. Persons are constantly attacked with convulsions—children especially; convulsions are spasms: spasms, of course, are to be treated by antispasmodics. This is the notion amongst people ignorant of medicine: it is the notion amongst old medical men: it is the notion amongst half educated young ones. All this time these convulsions are merely a symptom; that symptom depends upon, and denotes, most important disease in the brain: the only chance of saving life, is the prompt and vigorous application of proper remedies to the brain; but the practitioner whose mind is occupied with the symptom, and who prescribes antispasmodics, not only loses the time in which alone any thing can be done to snatch the victim from death, but by his remedies absolutely adds fuel to the flame which is consuming his patient. In disease of the hip-joint pain is felt, not in the hip, but, in the early stage of the disease, at the knee. This also depends on nervous communication. The most dreadful consequences daily occur from an ignorance of this single fact. In all these cases error is inevitable, without a knowledge of[Pg 7] anatomy: it is scarcely possible with it: in all these cases error is fatal: in all these cases anatomy alone can prevent the error—anatomy alone can correct it. Experience, so far from leading to its detection, would only establish it in men's minds, and render its removal impossible. What is called experience is of no manner of use to an ignorant and unreflecting practitioner. In nothing does the adage, that it is the wise only who profit by experience, receive so complete an illustration as in medicine. A man who is ignorant of certain principles, and who is incapable of reasoning in a certain manner, may have daily before him for fifty years cases affording the most complete evidence of their truth, and of the importance of the deduction to which they lead, without observing the one, or deducing the other. Hence the most profoundly ignorant of medicine, are often the oldest members of the profession, and those who have had the most extensive practice. A medical education, founded on a knowledge of anatomy, is, therefore, not only indispensable to prevent the most fatal errors, but to enable a person to obtain advantage from those sources of improvement which extensive practice may open to him.

To the surgeon, anatomy is eminently what Bacon has so beautifully said that knowledge in general is: it is power—it is power to lessen pain, to save life, and to eradicate diseases, which, without its aid, would be incurable and fatal.

It is impossible to convey to the reader a clear conception of this truth, without a reference to particular cases; and the subject is one of such extreme importance, that it may be worth while to direct the attention for a moment to two or three of the capital diseases which the surgeon is daily called upon to treat. Aneurism, for example, is a disease of an artery, and consists of a preternatural dilatation of its coats. This dilatation arises from the debility of the vessel, whence, unable to resist the impetus of the blood, it yields, and is dilated into a sac. When once the disease is induced, it commonly goes on to increase with a steady and uninterrupted progress, until at last it suddenly bursts, and the patient expires instantaneously from loss of blood. When left to itself, it almost uniformly proves fatal in this manner; yet, before the time of Galen, no notice was taken of this terrible malady. The ancients, indeed, who believed that the arteries were air tubes, could not possibly have conceived the existence of an aneurism. Were the number of individuals in Europe, who are now annually cured of aneurism, by the interference of art, to be assumed as the basis of a calculation of the number of persons who must have perished by this disease, from the beginning of the world to the time of Galen, it would convey some [Pg 8]conception of the extent to which anatomical knowledge is the means of saving human life.

The only way in which it is possible to cure this disease is, to produce an obliteration of the cavity of the artery. This is the object of the operation. The diseased artery is exposed, and a ligature is passed around it, above the dilatation, by means of which the blood is prevented from flowing into the sac, and inflammation is excited in the vessel; in consequence of which its sides adhere together, and its cavity becomes obliterated. The success of the operation depends entirely on the completeness of the adhesion of the sides of the vessel, and the consequent obliteration of its cavity. This adhesion will not take place unless the portion of the artery to which the ligature is applied be in a sound state. If it be diseased, as it almost always is near the seat of the aneurism, when the process of nature is completed by which the ligature is removed, hemorrhage takes place, and the patient dies just as if the aneurism had been left to itself. For a long time the ligature was applied as close as possible to the seat of the aneurism: the aneurismal sac was laid open in its whole extent, and the blood it contained was scooped out. The consequence was, that a large deep-seated sore, composed of parts in an unhealthy state, was formed: it was necessary to the cure that this sore should suppurate, granulate, and heal: a process which the constitution was frequently unable to support. Moreover, there was a constant danger that the patient would perish from hemorrhage, through the want of adhesion of the sides of the artery. The profound knowledge of healthy and of diseased structure, and of the laws of the animal economy by which both are regulated, which John Hunter had acquired from anatomy, suggested to this eminent man a mode of operating, the effect of which, in preserving human life, has placed him high in the rank of the benefactors of his race. This consummate anatomist saw, that the reason why death so often followed the common operation was, because that process which was essential to his success was prevented by the diseased condition of the artery. He perceived that the vessel, at some distance from the aneurism, was in a sound state; and conceived, that if the ligature were applied to this distant part, that is, to a sound instead of a diseased portion of the artery, this necessary process would not be counteracted. To this there was one capital objection, that it would often be necessary to apply the ligature around the main trunk of an artery, before it gives off its branches, in consequence of which the parts below the ligature would be deprived of their supply of blood, and would therefore mortify. So frequent and great are the communications between all the arteries of the body, however, that he thought[Pg 9] it probable, that a sufficient supply would be borne to these parts through the medium of collateral branches. For an aneurism in the ham, he, therefore, boldly cut down upon the main trunk of the artery which supplies the lower extremity; and applied a ligature around it, where it is seated near the middle of the thigh, in the confident expectation that, though he thus deprived the limb of the supply of blood which it received through its direct channel, it would not perish. His knowledge of the processes of the animal economy, led him to expect that the force of the circulation being thus taken off from the aneurismal sac, the progress of the disease would be stopped; that the sac itself, with all its contents, would be absorbed; that by this means the whole tumor would be removed, and that an opening into it would be unnecessary. The most complete success followed this noble experiment, and the sensations which this philosopher experienced when he witnessed the event, must have been exquisite, and have constituted an appropriate reward for the application of profound knowledge to the mitigation of human suffering. After Hunter followed Abernethy, who, treading in the footsteps of his master, for an aneurism of the femoral, placed a ligature around the external iliac artery; lately the internal iliac itself has been taken up, and surgeons have tied arteries of such importance, that they have been themselves astonished at the extent and splendor of their success. Every individual, on whom an operation of this kind has been successfully performed, is snatched by it from certain and inevitable death!

The symptom by which an aneurism is distinguished from every other tumor is, chiefly its pulsating motion. But when an aneurism has become very large, it ceases to pulsate; and when an abscess is seated near an artery of great magnitude, it acquires a pulsating motion; because the pulsations of the artery are perceptible through the abscess. The real nature of cases of this kind cannot possibly be ascertained, without a most careful investigation, combined with an exact knowledge of the structure and relative position of all the parts in the neighborhood of the tumor. Pelletan, one of the most distinguished surgeons of France, was one day called to a man who, after a long walk, was seized with a severe pain in the leg, over the seat of which appeared a tumor, which was attended with a pulsation so violent that it lifted up the hand of the examiner. There seemed every reason to suppose that the case was an aneurismal swelling. This acute observer, however, in comparing the affected with the sound limb, perceived in the latter a similar throbbing. On careful examination he discovered that, by a particular disposition in this individual, one of the main arteries of the leg (the anterior tibial) deviated from[Pg 10] its usual course, and instead of plunging deep between the muscles, lay immediately under the skin and fascia. The truth was, that the man in the exertion of walking, had ruptured some muscular fibres, and the uncommon distribution of the artery gave to this accident these peculiar symptoms. The real nature of this case could not possibly have been ascertained but by an anatomist. The same surgeon has recorded the case of a man who, having fallen twice from his horse, and experienced for several years considerable uneasiness in his back, was afflicted with acute pain in the abdomen. At the same time an oval, irregularly circumscribed tumor made its appearance in the right flank. It presented a distinct fluctuation, and had all the appearance of a collection of matter depending on caries of the vertebræ. The pain was seated chiefly at the lower portion of that part of the spine which forms the back, which was, moreover, distorted; and this might have confirmed the opinion that the case was a lumbar abscess with caries. Pelletan, however, who well knew that an aneurism, as it enlarges, may destroy any bone in its neighborhood, saw that the disease was an aneurism, and predicted that the patient must perish. On opening the body (for the man lived only ten days after Pelletan first saw him) an aneurismal tumor was discovered, which nearly filled the cavity of the abdomen. If this case had been mistaken for lumbar abscess, and the tumor had been opened with a view of affording an exit to the matter, the man would have died in a few seconds. There is no surgeon of discernment or experience whose attention has not been awakened, and whose sagacity has not been put to the test, by the occurrence of similar cases in his own practice. The consequence of error is almost always instantaneously fatal. The catalogue of such disastrous events is long and melancholy. Richerand has recorded, that Ferrand, head surgeon of the Hotel Dieu, mistook an aneurism in the armpit for an abscess; plunged his knife into the swelling, and killed the patient. De Haen speaks of a person who died in consequence of an opening which was made, contrary to the advice of Boerhaave in a similar tumor at the knee. Vesalius was consulted about a tumor in the back, which he pronounced to be an aneurism; but an ignorant practitioner having made an opening into it, the patient instantly bled to death. Nothing can be more easy than to confound an aneurism of the artery of the neck with the swelling of the glands in its neighborhood: with a swelling of the cellular substance which surrounds the artery; with abscesses of various kinds; but if a surgeon were to fall into this error, and to open a carotid aneurism, his patient would certainly be dead in the space of a few moments. It must be evident, then, that a thorough knowledge[Pg 11] of anatomy is not only indispensable to the proper treatment of cases of this description, but also to the prevention of the most fatal mistakes.

There is nothing in surgery of more importance than the proper treatment of hemorrhage. Of the confusion and terror occasioned by the sight of a human being from whom the blood is gushing in torrents, and whose condition none of the spectators is able to relieve, no one can form an adequate conception, but those who have witnessed it. In all such cases, there is one thing proper to be done, the prompt performance of which is generally as certainly successful, as the neglect of it is inevitably fatal. It is impossible to conceive of a more terrible situation than that of a medical man who knows not what to do on such an emergency. He is confused; he hesitates: while he is deciding what measures to adopt, the patient expires: he can never think of that man's death without horror, for he is conscious that, but for his ignorance, he might have averted his patient's fate. The ancient surgeons were constantly placed in this situation, and the dread inspired by it retarded the progress of surgery more than all other causes put together. Not only were they terrified from interfering with the most painful and destructive diseases, which experience has proved to be capable of safe and easy removal, but they were afraid to cut even the most trivial tumor. When they ventured to remove a part, they attempted it only by means of the ligature, or by the application of burning irons. When they determined to amputate, they never thought of doing so until the limb had mortified, and the dead had separated from the living parts; for they were absolutely afraid to cut into the living flesh. They had no means of stopping hemorrhage, but by the application of astringents to the bleeding vessels, remedies which were inert; or of burning irons, or boiling turpentine, expedients which were not only inert but cruel. Surgeons now know that the grand means of stopping hemorrhage is compression of the bleeding vessel. If pressure be made on the trunk of an artery, though blood be flowing from a thousand branches given off from it, the bleeding will cease. Should the situation of the artery be such as to allow of effectual external pressure, nothing further is requisite: the pressure being applied, the bleeding is stopped at once: should the situation of the vessel place it beyond the reach of external pressure, it is necessary to cut down upon it, and to secure it by the application of a ligature. Parè may be pardoned for supposing that he was led to the discovery of this invaluable remedy by the inspiration of the Deity. By means of it the most formidable operations may be undertaken with the utmost confidence, because the wounded vessels can be secured the moment they are cut:[Pg 12] by the same means the most frightful hemorrhages may be most effectually stopped: and even when the bleeding is so violent as to threaten immediate death, it may often be averted by the simple expedient of placing the finger upon the wounded vessel, until there is time to tie it. But it is obvious that none of these expedients can be employed, and that these bleedings can neither be checked at the moment, nor permanently stopped, without such a knowledge of the course of the trunks and branches of vessels, as can be acquired only by the study of anatomy.

The success of amputation is closely connected with the knowledge of the means of stopping hemorrhage. Not to amputate is often to abandon the patient to a certain and miserable death. And all that the surgeon formerly did, was to watch the progress of that death: he had no power to stop or even to retard it. The fate of Sir Philip Sidney is a melancholy illustration of this truth. This noble minded man, the light and glory of his age, was cut off in the bloom of manhood, and the midst of his usefulness, by the wound of a musket bullet in his left leg, a little above the knee, "when extraction of the ball, or amputation of the limb," says his biographer, "would have saved his inestimable life: but the surgeons and physicians were unwilling to practice the one, and knew not how to perform the other. He was variously tormented by a number of surgeons and physicians for three weeks." Amputation indeed was never attempted, except where mortification had itself half performed the operation. The just apprehension of an hemorrhage which there was no adequate means of stopping, checked the hand of the boldest surgeon, and quailed the courage of the most daring patient—and if ever the operation was resorted to, it almost always proved fatal: the patient generally expired, according to the expression of Celsus, "*in ipso opere.*" How could it be otherwise? The surgeon cut through the flesh of his patient with a red hot knife: this was his only means of stopping the hemorrhage: by this expedient he sought to convert the whole surface of the stump into an eschar: but this operation, painful in its execution, and terrible in its consequences, when it even appeared to succeed, succeeded only for a few days; for the bleeding generally returned, and proved fatal as soon as the sloughs or dead parts became loose. Plunging the stump into boiling oil, into boiling turpentine, into boiling pitch, for all these means were used, was attended with no happier result, and after unspeakable suffering, almost every patient perished. In the manner in which amputation is performed at present, not more than one person in twenty loses his life in consequence of the operation, even taking into the account all the cases in which it is practised in hospitals. In [Pg 13]private practice, where many circumstances favor its success, it is computed that 95 persons out of 100 recover from it, when it is performed at a proper time, and in a proper manner. It seems impossible to exhibit a more striking illustration of the great value of anatomical knowledge.

But if there be any disease, which, from the frequency of its occurrence, from the variety of its forms, from the difficulty of discriminating between it and other maladies, and from the danger attendant on almost all its varieties, requires a combination of the most minute investigation, with the most accurate anatomical knowledge, it is that of hernia. This disease consists of a protrusion of some of the viscera of the abdomen, from the cavity in which they are naturally contained, into a preternatural bag, composed of the portion of the peritoneum (the membrane which lines the abdomen) which is pushed before them. It is computed that one sixteenth of the human race are afflicted with this malady. It is sometimes merely an inconvenient complaint, attended with no evil consequences whatever; but there is no form of this disease, which is not liable to be suddenly changed, and by slight causes, from a perfectly innocent state, into a condition which may prove fatal in a few hours. The disease itself occurs in numerous situations; it may be confounded with various diseases; it may exist in the most diversified states; it may require, without the loss of a single moment, a most important and delicate operation; and it may appear to demand this operation, while the performance of it may really be not only useless, but highly pernicious.

The danger of hernia depends on its passing into that state which is technically termed strangulation. When a protruded intestine suffers such a degree of pressure, as to occasion a total obstruction to the passage of its contents, it is said to be strangulated. The consequence of pressure thus producing strangulation is, the excitement of inflammation: this inflammation must inevitably prove fatal, unless the pressure be promptly removed. In most cases, this can be effected only by the operation. Two things, then, are indispensable: first, the ability to ascertain that the symptoms are really produced by pressure, that is, to distinguish the disease from the affections which resemble it; and secondly, when this is effected, to perform the operation with promptitude and success. The distinction of strangulated hernia from affections which resemble it, often requires the most exact knowledge and the most minute investigation. The intestine included in a hernial sac, may be merely affected with colic, and thus give rise to the appearance of strangulation. It may be in a state of irritation, produced, for example, by unusual fatigue; and from[Pg 14] this cause, may be attacked with the symptoms of inflammation. Inflammation may be excited in the intestine, by the common causes of inflammation, which the hernia may have no share in inducing, and of which it may not even participate. Were this case mistaken, and the operation performed, it would not only be useless, but pernicious: while the attention of the practitioner would be diverted from the real nature of the malady; the prompt and vigorous application of the remedies which alone could save the patient, would be neglected, and he would probably perish. On the other hand, a very small portion of intestine may become strangulated, and urgently require the operation. But there may be no tumor; all the symptoms may be those, and, on a superficial examination, only those, of inflammation of the bowels. Were the real nature of this case mistaken, death would be inevitable. Nothing is more common than fatal errors of this kind. It is only a few months ago, that a physician was called in haste to a person who was said to be dying of inflammation of the bowels. Before he reached the house the man was dead. He had been ill only three days. On looking at the abdomen, there was a manifest hernia: the first glance was sufficient to ascertain the fact. The practitioner in attendance had known nothing of the matter; he had never suspected the real nature of the disease, and had made no inquiry which could have led to the detection of it. Here was a case which might probably have been saved, but for the criminal ignorance and inattention of the practitioner. Whenever there are symptoms of inflammation of the bowels, examination of the abdomen is indispensable: and the life of the patient will depend on the care and accuracy with which the investigation is made.

But it is possible that inflammation may attack the parts included in the hernial sac, without arising from the hernia itself. The inflammation may be produced by the common causes of inflammation; there may be no pressure: there may be no strangulation: the swelling may be the seat, not the cause of the disease. In this case, too, the operation would be both useless and pernicious. Now all these are diversities which it is of the highest importance to discriminate. In some of them, life depends on the clearness, accuracy, and promptitude, with which the discrimination is made. Promptitude is of no less consequence than accuracy. If the decision be not formed and acted on at once, it will be of no avail. The rapidity of the progress of this disease is often frightful. We have mentioned a case in which it was fatal in three days, but it not unfrequently terminates fatally in less than twenty four hours. Sir Astley Cooper mentions a case in which the patient was dead in[Pg 15] eight hours after the commencement of the disease. Larrey has recorded the case of a soldier in whom a hernia took place, which was strangulated immediately. He was brought to the "ambulance" instantly, and perished in two hours with gangrene of the part, and of the abdominal viscera. This was the second instance which had occurred to this surgeon of a rapidity thus appalling. What clearness of judgment, what accuracy of knowledge, what promptitude of decision, are necessary to treat such a disease with any chance of success!

The moment that a case is ascertained to be strangulated hernia, an attempt must be made to liberate the parts from the stricture, and to replace them in their natural situation. This is first attempted by the hand, and the operation is technically termed the *taxis*. The patient must be placed in a particular position; pressure must be made in a particular direction; it is impossible to ascertain either, without an accurate knowledge of the parts. If pressure be made in a wrong direction, and in a rough and unscientific manner, the organs protruded instead of being urged through a proper opening, are bruised against the parts which oppose their return. Many cases are on record, in which gangrene and even rupture of the intestine, have been occasioned in this manner. When the parts cannot be returned by the hand, assisted by those remedies which experience has proved to be beneficial, the operation must be performed without the delay of a moment. To its proper performance two things are necessary. First, a minute anatomical knowledge of the various and complicated parts which are implicated in it; and secondly, a steady, firm, and delicate command of the knife. In the first place, the integuments must be divided; the cellular substance which intervenes between the skin and the hernial sac must be removed layer by layer with the knife and the dissecting forceps; the sac itself must be opened: this part of the operation must be performed with the most extreme caution: the sac being laid open, the protruded organs are now exposed to view. The operator must next ascertain the exact point where the stricture exists; having discovered its seat, he must make his incision with a particular instrument— in a certain direction—to a definite extent. On account of the nature of the parts implicated in the operation, and the proximity of vessels, life depends on an exact knowledge and a precise and delicate attention to all these circumstances. How can this knowledge be obtained, how can this dexterity be acquired, without a profound acquaintance with anatomy, and how can this be acquired without frequent and laborious dissection? The eye must become familiar with the appearance of the integuments, with the appearance of the cellular [Pg 16]substance beneath it, with the appearance of the hernial sac, and of the changes which it undergoes by disease; with the appearance of the various viscera contained in it, and of their changes: and the hand must pay that steady and prompt obedience to the judgment, which nothing but knowledge, and the consciousness of knowledge, can command. Even this is not all. When the operation has been performed thus far with perfect skill and success, the most opposite measures are required according to the actual state of the organs contained in the sac. If they are agglutinated together—if portions of them are in a state of mortification, to return them into the cavity of the abdomen in that condition, would, in general, be certain death. Preternatural adhesion must be removed; mortified portions must be cut away: but how can this possibly be done without an acquaintance with healthy and diseased structure, and how can this be obtained without dissecting the organs in a state of health and of disease?

It has been stated that the progress of strangulated hernia to a fatal termination is often frightfully rapid; in certain cases to delay the operation, even for a very short period, is, therefore, to lose the only chance of success. But ignorant and half informed surgeons are afraid to operate. They are conscious that the operation is one of immense importance: they know that in the hands of an operator ignorant of anatomy, it is one of extreme hazard: they therefore put off the time as long as possible: they have recourse to every expedient: they resort to every thing but the only efficient remedy, and when at last they are compelled by a secret sense of shame to try that, it is too late. All the best practical surgeons express themselves in the strongest language on the importance of performing the operation early, if it be performed at all. On this point there is a perfect accordance between the most celebrated practitioners on the continent, and the great surgeons of our own country: all represent, in many parts of their writings, the dangerous and fatal effects of delay. Mr. Hey in his Practical Observations, states that when he first began to practice, he considered the operation as the last resource, and only to be employed when the danger appeared imminent. "By this dilatory mode of practice," says he, "I lost three patients in five, upon whom the operation was performed. Having more experience of the urgency of the disease, I made it my custom, when called to a patient who had laboured two or three days under the disease, to wait only about two hours, that I might try the effect of bleeding (if that evacuation was not forbidden by some peculiar circumstance of the case) and the tobacco clyster. In this mode of practice, I lost about two patients in nine, upon whom I[Pg 17] operated. This comparison is drawn from cases nearly similar, leaving out of the account those cases in which gangrene of the intestine had taken place. I have now, at the time of writing this, performed the operation thirty-five times; and have often had occasion to lament that I performed it too late, but never that I had performed it too soon."

These observations are sufficient to show the importance of anatomy in certain surgical diseases. The state of medical opinion from the earliest ages to the present time, furnishes a most instructive proof of its necessity to the detection and cure of disease in general. The doctrines of the father of physic were in the highest degree vague and unmeaning. Every thing is resolved by Hippocrates into a general principle, which he terms nature; and to which he ascribes intelligence; which he clothes with the attributes of justice; and which he represents as possessing virtues and powers, which he says are her servants, and by means of which she performs all her operations in the bodies of animals, distributes the blood, spirits, and heat, through all the parts of the body, and imparts to them life and sensation. He states that the manner in which she acts, is by attracting what is good or agreeable to each species, and retaining, preparing, and changing it: or, on the other hand, by rejecting whatever is superfluous or hurtful, after she has separated it from the good. This is the foundation of the doctrine of depuration, concoction, and crisis in fevers, so much insisted on by him, and by other physicians after him; but when he explains what he means by nature, he resolves it into heat, which he says appears to have something immortal in it.

The great opponent of Hippocrates was Asclepiades. He asserted that matter, considered in itself, is of an unchangeable nature: that all perceptible bodies are composed of a number of small ones, termed corpuscles, between which there are interspersed an infinity of small spaces totally devoid of matter: that the soul itself is composed of these corpuscles: that what is called nature is nothing more than matter and motion: that Hippocrates knew not what he said when he spoke of nature as an intelligent being, and ascribed to her various qualities and virtues: that the corpuscles, of which all bodies are composed, are of different figures, and consist of different assemblages: that all bodies contain numerous pores, or interstices, which are of different sizes: that the human body, like all other bodies, possesses pores peculiar to itself: that these pores are larger or smaller, according as the corpuscles which pass through them differ in magnitude: that the blood consists of the largest, and the spirits and the heat of the smallest.[Pg 18] On these principles, Asclepiades founded his theory of medicine. He maintains, that as long as the corpuscles are freely received by the pores, the body remains in its natural state: that, on the contrary, as soon as any obstacle obstructs their passage, it begins to recede from that state: that, therefore, health depends on the just proportion between these pores and corpuscles: that, on the contrary, disease proceeds from a disproportion between them: that the most usual obstacle arises from a retention of some of the corpuscles in their ordinary passages, where they arrive in too large a number, or are of irregular figures, or move too fast or proceed too slow: that phrensies, lethargies, pleurises, burning fevers for example, are occasioned by these corpuscles stopping of their own accord: that pain is produced by the stagnation of the largest of all these corpuscles, of which the blood consists: that, on the contrary, deliriums, languors, extenuations, leanness and dropsies, derive their origin from a bad state of the pores, which are too much relaxed, or opened: that dropsy, in particular, proceeds from the flesh being perforated with various small holes, which convert the nourishment received into them into water: that hunger is occasioned by an opening of the large pores of the stomach and belly: that thirst arises from an opening of the small pores: that intermittent fevers have the same origin: that quotidian fever is produced by a retention of the largest corpuscles; tertian fever by a retention of corpuscles somewhat smaller; and quartan fever by a retention of the smallest corpuscles of all.

Galen maintained that the animal body is composed of three principles, namely, the solids, the humors, and the spirits. That the solid parts consist of similar and organic: that the humors are four in number, namely, the blood, the phlegm, the yellow bile, and the black bile: that the spirits are of three kinds, namely, the vital, the animal, and the natural: that the vital spirit is a subtle vapour which arises from the blood, and which derives its origin from the liver, the organ of sanguification: that the spirits thus formed, are conveyed to the heart, where, in conjunction with the air drawn into the lungs by respiration, they become the matter of the second species, namely, of the vital spirits: that in their turn, the vital spirits are changed into the animal in the brain, and so on.

At last came Paracelsus, who was believed to have discovered the elixir of life, and who is the very prince of charlatans. He delivered a course of lectures on the theory and practice of physic in the University of Basle, which he commenced by burning the works of Galen and Avicenna in the presence of his auditory. He assured his hearers, that his shoe-latchets had more knowledge than both these[Pg 19] illustrious authors put together: that all the academies in the world had not so much experience as his beard; and that the hair on the back of his neck was more learned than the whole tribe of authors. It was fitting that a person of such splendid pretensions should have a magnificent name. He, therefore, called himself PHILIPPUS AUREOLUS THEOPHRASTUS PARACELSUS BOMBAST VON HOHENHEIM. He was a great chemist, and like other chemists, he was a little too apt to carry into other sciences "the smoke and tarnish of the furnace." He conceived that the elements of the living system were the same as those of his laboratory, and that sulphur, salt, and quicksilver, were the constituents of organized bodies. He taught that these constituents were combined by chemical operations: that their relations were governed by Archeus, a demon, who performed the part of alchemist in the stomach, who separated the poisonous from the nutritive part of the food, and who communicated the tincture by which the food became capable of assimilation: that this governor of the stomach, this *spiritus vitæ*, this astral body of man, was the immediate cause of all diseases, and chief agent in their cure: that each member of the body had its peculiar stomach, by which the work of secretion was effected: that diseases were produced by certain influences, of which there were five in particular, viz. *ens estrale, ens veneni, ens naturale, ens spirituale*, and *ens deale*: that when Archeus was sick, putrescence was occasioned, and that either *localiter* or *emunctorialiter*, &c. &c. &c.

It would be leading to a detail which is incompatible with our present purpose to follow these speculations, or to give an account of the doctrines of the mechanical physicians, who believed that every operation of the animal economy was explained by comparing it to a system of ropes, levers, and pulleys, united with a number of rigid tubes of different lengths and diameters, containing fluids which, from variations in their impelling causes, moved with different degrees of velocity: or of the chemical physicians, whose manner of theorizing and investigating would have qualified them better for the occupation of the brewer or of the distiller, than for that of the physician. All these speculations are idle fancies, without any evidence whatever to support them; and it has been argued that, for this very reason, they must have been without any practical result, and that, therefore, if they were productive of no benefit, they were, at least, innoxious. No opinion can be more false or pernicious. These wretched theories not only pre occupied the mind, prevented it from observing the real phenomena of health and of disease, and the actual effect of the remedies which were employed, and thus put an effectual stop to[Pg 20] the progress of the science: but they were productive of the most direct and serious evils. It is no less true in medicine than in philosophy and morals, that there is no such thing as innoxious error; that men's opinions invariably influence their conduct; and that physicians, like other men, act as they think. Asclepiades, whose mind was full of corpuscles and interstices, was intent on finding suitable remedies, which he discovered in gestation, friction, and the use of wine. By various exercises, he proposed to render the pores more open, and to make the juices and corpuscles, the retention of which causes disease, to pass more freely. Hence he used gestation from the very beginning of the most burning fevers. He laid it down as a maxim, that one fever was to be cured by another; that the strength of the patient was to be exhausted by making him watch and endure thirst to such a degree, that for the first two days of the disorder he would not allow them to cool their mouths with a drop of water. Abernethy's regulated diet is luxurious compared to his plan of abstinence. For the three first days he allowed his patients no aliment whatever; on the fourth, he so far relented as to give to some of them a small portion of food; but from others he absolutely withheld all nourishment till the seventh day. And this is the gentleman who laid it down as a maxim, that all diseases are to be cured "*Tuto, celeriter et jucunde.*" To be sure he was a believer in the doctrine of compensation; and in the latter stage of their diseases endeavored to recompense his patients for the privations he caused them to endure in the beginning of their illness. Celsus observes, that though he treated his patients like a butcher during the first days of the disorder, he afterwards indulged them so far as to give directions for making their beds in the softest manner. He allowed them abundance of wine, which he gave freely in all fevers; he did not forbid it even to those afflicted with phrenzy: nay, he ordered them to drink it till they were intoxicated; for, said he, it is absolutely necessary that persons who labor under phrenzy should sleep, and wine has a narcotic quality. To lethargic patients, he prescribed it with great freedom, but with the opposite purpose of rousing them from their stupor. His great remedy in dropsy was friction, which, of course, he employed to open the pores. With the same view, he enjoined active exercise to the sick; but what is a little extraordinary, he denied it to those in health.

Eristratus, who was a great speculator, and whose theories had the most important influence on his practice, banished blood-letting altogether from medicine, for the following notable reasons: because, he says, we cannot always see the vein we intend to open; because we are not sure we[Pg 21] may not open an artery instead of a vein; because we cannot ascertain the true quantity to be taken; because, if we take too little, the intention is not answered; if too much, we may destroy the patient; and because the evacuation of the venous blood is succeeded by that of the spirits, which thus pass from the arteries into the veins; wherefore, blood-letting ought never to be used as a remedy in disease. Yet, though he was thus cautious in abstracting blood, it must not be supposed that he was not a sufficiently bold practitioner. In tumor of the liver, he hesitated not to cut open the abdomen, and to apply his medicines immediately to the diseased organ; but though he took such liberties with the liver, he regarded with the greatest apprehension the operation of tapping in dropsy of the abdomen: because, said he, the waters being evacuated, the liver which is inflamed and become hard like a stone, is more pressed by the adjacent parts, which the waters kept at a distance from it, whence the patient dies.

One physician conceived that gout originated from an effervescence of the synovia of the joints with the vitriolated blood: whence he recommended alcohol for its cure: a remedy for which the court of aldermen ought to have voted him a medal. A more ancient practitioner, who believed that the finger of St. Blasius was very efficacious "for removing a bone which sticks in the throat," maintained that gout was the "grand drier," and prescribed a remedy for it, which the patient was to use for a whole year, and to observe the following diet each month. In September, he must eat and drink milk; in October, he must eat garlic; in November, he is to abstain from bathing; in December, he must eat no cabbage; in January, he is to take a glass of pure wine in the morning; in February, to eat no beef; in March, to mix several things both in eatables and drinkables; in April, not to eat horse-radish; nor in May, the fish called Polypus; in June, he is to drink cold water in a morning; in July, to avoid venery; and lastly, in August, to eat no mallows.

A third physician deduced all diseases from inspissation of the fluids; hence he attached the highest importance to diluent drinks, and believed that tea, especially, is a sovereign remedy in almost every disease to which the human frame is subject; "tea," says Bentekoe, who is loudest in his praises of this panacea, and who, as Blumenbach observes, 'deserved to have been pensioned by the East India Company for his services,' "tea is the best, nay, the only remedy for correcting viscidity of the blood, the source of all diseases, and for dissipating the acid of the stomach, as it contains a fine oleaginous volatile salt, and certain subtle spirits which are analogous in their nature to the animal[Pg 22] spirits. Tea fortifies the memory and all the intellectual faculties: it will therefore furnish the most effectual means of improving physical education. Against fever there is no better remedy than forty or fifty cups of tea, swallowed immediately after one another, the slime of the pancreas is thus carried off."

Another physician derived all his diseases from a redundancy or deficiency of fire and water. He maintained that where the water predominated, the fluids became viscid, and that hence arose intermittent fevers and arthritic complaints. His remedies are in strict conformity to his theory. These diseases are to be cured by volatile salts, which abound with fiery particles; venesection in any case is highly pernicious; these fiery medicines are the only efficacious remedies, and are to be employed even in diseases of the most inflammatory nature. "Life," says Dr. Brown, "is a forced state;" it is a flame kept alive by excitement; every thing stimulates; some substances too violently; others not sufficiently; there are thus two kinds of debility, indirect and direct, and to one or other of these causes must be referred the origin of all diseases. According to this doctrine, the mode of cure is simple: we have nothing to do but to supply, to moderate or to abstract stimuli. Typhus fever, in this system, is a disease of extreme debility; we must therefore give the strongest stimulants. Consumption and apoplexy, also, are diseases of debility; of course, the remedies are active stimulants. Humanity shudders, and with reason, at the application of such doctrines to practice. And not less destitute of reason, and not less dangerous in practice, is the great doctrine of debility promulgated by Cullen. This celebrated professor taught, that the circumstance which invariably characterised fever, that which constituted its essence, was debility. The inference was obvious, that, above all things, the strength must be supported. The consequence was, that blood-letting was neglected, and that bark and wine were given in immense quantities, in cases in which intense inflammation existed. The practice was in the highest degree mortal; the number of persons who have perished in consequence of this doctrine is incalculable. So far then is it from being true, that medical theories are of no practical importance, there is the closest possible connection between the speculations of the physician in his closet, and the measures which he adopts at the bed side of his patient. Truth to him is a benignant power, which stops the progress of disease, protracts the duration of life, and mitigates the suffering it may be unable to remove: error is a fearfully active and tremendously potent principle. There is not a medical prejudice which has not slain its thousands, nor a false theory which[Pg 23] has not immolated its tens of thousands. The system of medicine and surgery which is established in any country, has a greater influence over the lives of its inhabitants than the epidemic diseases produced by its climate, or the decisions of its government concerning peace and war. The devastations of the yellow fever will bear no comparison with the ravages committed by the Brunonian system; and the slaughter of the field of Waterloo counts not of victims, a tithe of the number of which the Cullenian doctrine of debility can justly boast. Anatomy alone will not teach a physician to think, much less to think justly; but it will give him the elements of thinking; it will furnish him with the means of correcting his errors; it will certainly save him from some delusions, and will afford to the public the best shield against his ignorance, which may be fatal, and against his presumption, which may be devastating.

We have entered into this minute detail at the hazard, we are aware, of tiring the reader; but in the hope of leaving on his mind a more distinct impression of the importance of anatomical knowledge, than could possibly be produced by a mere allusion to the circumstances which have been explained. In all ages, formidable obstacles have opposed the prosecution of anatomical investigations. Among these, without doubt, the most powerful has its source in a feeling which is natural to the heart of man. The sweetest, the most sacred associations are indissolubly connected with the person of those we love. It is with the corporeal frame that our senses have been familiar: it is that on which we have gazed with rapture; it is that which has so often been the medium of conveying to our hearts the thrill of exstacy. We cannot separate the idea of the peculiarities and actions of a friend from the idea of his person. It is for this reason that "every thing which has been associated with him acquires a value from that consideration; his ring, his watch, his books, and his habitation. The value of these as having been his, is not merely fictitious; they have an empire over my mind; they can make me happy or unhappy; they can torture and they can tranquilize; they can purify my sentiments, and make me similar to the man I love; they possess the virtue which the Indian is said to attribute to the spoils of him he kills, and inspire me with the power, the feelings, and the heart of their preceding master." It is nothing, says the survivor, to tell me, when disease completed its work and death has seized its prey, that that body, with which are connected so many delightful sensations, is a senseless mass of matter: that it is no longer my friend; that the spirit which animated it, and rendered it lovely to my sight and dear to my affections, is gone. I know that it is gone, I know that I never more shall see the light of [Pg 24]intelligence brighten that countenance, nor benevolence beam in that eye, nor the voice of affection sound from those lips: that which I loved, and which loved me, is not here: but here are still the features of my friend: this is his form, and the very particles of matter which compose this dull mass, a few hours ago were a real part of him, and I cannot separate them, in my imagination, from him. And I approach them with the profounder reverence; I gaze upon them with the deeper affection, because they are all that remain to me. I would give all that I possess to purchase the art of preserving the wholesome character and rosy hue of this form, that it might be my companion still: but this is impossible: I cannot detain it from the tomb: but when I have "cast a heap of mould upon the person of my friend, and taken the cold earth for its keeper," I visit the spot in which it is deposited with awe: it is sacred to my imagination: it is dear to my heart. There is a real and deep foundation for these feelings in human nature: they arise spontaneously in the bosom of man, and we see their expression and their power in the customs of all nations, savage as well as civilized, and in the conduct of all men, the most ignorant and uncultivated no less than the most intelligent and refined. It has been the policy of society to foster these sentiments. If has been conceived that the sanctity which attaches to the dead, is reflected back in a profounder feeling of respect for the living; that the solemnity with which death is regarded, elevates, in the general estimation, the value of life; and that he who cannot approach the mortal remains of a fellow creature without an emotion of awe, must regard with horror every thing which places in danger the life of a human being. Religion has contributed indirectly, but powerfully, to the strength and perpetuity of these impressions; and superstition has availed herself of them to play her antics, and to accomplish her base and malignant purposes. It is not the eradication of these feelings that can be desired, but their control: it is not the extinction of these natural and useful emotions that is pleaded for, but they should give way to higher considerations when these exist. Veneration for the dead is connected with the noblest and sweetest sympathies of our nature: but the promotion of the happiness of the living is a duty from which we can never be exonerated.

In antient times the voice of reason could not be heard. Superstition, and customs founded on superstition, excited an influence which was neither to be resisted nor evaded. Dissection was then regarded with horror. In the warm countries of the east, the pursuit must have been highly offensive and even dangerous, and it was absolutely incompatible with the notions and ceremonies universally prevalent[Pg 25] in those days. The Jewish tenet of pollution must have formed an insuperable obstacle to the cultivation of anatomy amongst that people. By the Egyptians, every one who cut open a dead body was regarded with inexpressible horror. The Grecian philosophers so far overcame the prejudice, as occasionally to engage in the pursuit, and the first dissection on record was one made by Democritus of Abdera, the friend of Hippocrates, in order to discover the course of the bile. The Romans contributed nothing to the progress of the art: they were content with propitiating the Deities who presided over health and disease. They erected on the Palatine Mount a temple to the goddess Febris, whom they worshipped from a dread of her power. They also sacrificed to the goddess Ossipaga, who, it seems, presided over the growth of the bones, and to another styled Carna, who took care of the viscera, and to whom they offered bean broth and bacon, because these were the most nutritious articles of diet. The Arabians adopted the Jewish notion of pollution, and were thus prohibited by the tenets of their religion from practising dissection. Abdollaliph, who flourished about the year 1200, a man of learning and a teacher of anatomy, never saw and never thought of a human dissection. In order to examine and demonstrate the bones, he took his students to burying grounds, and earnestly recommended them, instead of reading books, to adopt that method of study: yet he seemed to have no conception that the dissection of a recent subject might be a still better method of learning. Christians were equally hostile to dissection. Pope Boniface the 8th issued a bull prohibiting even the maceration and preparation of skeletons. The priests were the only physicians, and so greatly did they abuse the office they assumed, that the evil at length became too intolerable to be borne. The church itself was obliged to prohibit the priesthood from interfering with the practice of medicine. All monks and canons who applied themselves to physic, were threatened with severe penalties, and all bishops, abbots, and priors who connived at their misconduct, were ordered to be suspended from their ecclesiastical functions. But it was not till three hundred years after this interdiction, that by a special bull which permitted physicians to marry, their complete separation from the clergy was effected.

In the 14th century, Mundinus, professor at Bologna, astonished the world by the public dissection of two human bodies. In the 15th century, Leonardo da Vinci contributed essentially to the progress of the art, by the introduction of anatomical plates, which were admirably executed. In the 16th century, the Emperor, Charles the 5th, ordered a consultation to be held by the divines of Salamanca, to [Pg 26]determine whether it was lawful, in point of conscience, to dissect a dead body in order to learn its structure. In the 17th century, Cortesius, professor of anatomy at Bologna, and afterwards professor of medicine at Messina, had long begun a treatise on practical anatomy, which he had an earnest desire to finish, but so great was the difficulty of prosecuting the study even in Italy, that in 24 years he could only twice procure an opportunity of dissecting a human body, and even then with difficulty and in hurry; whereas, he had expected to have done so, he says, once every year, according to the custom of the famous academies of Italy. In Muscovy, until very lately, both anatomy and the use of skeletons were positively forbidden; the first as inhuman, and the latter as subservient to witchcraft. Even the illustrious Luther was so biassed by the prejudices of his age, that he ascribed the majority of the diseases to the arts of the devil, and found great fault with physicians when they attempted to account for them by natural causes. England acquired the bad fame of being the country of witches, and opposed almost insuperable obstacles to the cultivation of anatomy. Even at present the prejudices of the people on this subject are violent and deeply rooted. The measure of that violence may be estimated by the degree of abhorrence with which they regard those persons who are employed to procure the subjects necessary for dissection. In this country, there is no other method of obtaining subjects but that of exhumation: aversion to this employment may be pardoned: dislike to the persons who engage in it is natural, but to regard them with detestation, to exult in their punishment, to determine for themselves its nature and measure, and to endeavor to assume the power of inflicting it with their own hands, is absurd. Magistrates have too often fostered the prejudices of the people, and afforded them the means of executing their vengeance on the objects of their aversion. The press has uniformly allied itself with the ignorance and violence of the vulgar, and has done every thing in its power to inflame the passions, which it was its duty to endeavor to soothe. It is notorious that the winter before last there was scarcely a week in which the papers did not contain the most exaggerated and disgusting statements: the appetite which could be gratified with such representations, was sufficiently degraded: but still more base was the servility which could pander to it. Half a century ago there was in Scotland no difficulty in obtaining the subjects which were necessary to supply the schools of anatomy. The consequence was, that medicine and surgery assumed new life—started from the torpor in which they had been spell-bound—and made an immediate, and rapid, and brilliant progress. The new [Pg 27]seminaries constantly sent into the world men of the most splendid abilities, at once demonstrating the excellence of the schools in which they were educated, and rendering them illustrious. Pupils flocked to them from all quarters of the globe, and they essentially contributed to that advancement of science which the present age has witnessed. In the 19th century, the good people of Scotland, that intelligent, that cool and calculating, that most reasonable and thinking people, have thought proper to return to the worst feeling and the worst conduct of the darkest periods of antiquity. There is at present no offence whatever, which seems to have such power to heat and exalt into a kind of torrent, the blood which usually flows so calmly and sluggishly in the veins of a Scotchman. The people of 1823 (to compare great things with small) emulate the spirit of those of their forefathers who "*were out in the forty-five*;" the object, to be sure, is somewhat different, but it is amusing to see the intensity and seriousness of the excitement. About twelve months ago an honest farmer of the name of Scott, who resides at Linlithgow, apprehended a poor wight who was pursuing his vocation, we presume, in the churchyard of that place; and this service appeared so meritorious to the people in his neighborhood, that they absolutely presented him with a piece of plate. In the winter sessions of 1822-3, a body was discovered on its way to the lecture-room of an anatomist in Glasgow, and in spite of the exertions of the police, aided by those of the military, this gentleman's premises and their contents, which were valuable, were entirely destroyed by the mob. For some time after this achievement, it was necessary to station a military guard at the houses of all the medical professors in that city. In the spring circuit of the justiciary court last year at Stirling, while the judges were proceeding to the court, the procession was assaulted with missiles; several persons were injured, and it was necessary to call in the protection of a military force. The object of the mob was to inflict summary punishment on a man who was about to be tried for the exhumation of a body. We happen to know that the most disgraceful proceedings were some time ago instituted in that town against a young gentleman of respectable family and connections, who was in fact expatriated, and whose prospects in life were entirely changed, if not ruined, because he had too much honor to implicate his instructors in a transaction which would have put them to an inconvenience, and in which they had engaged from a desire faithfully to discharge their duty to their pupils. Within the last five years three men were lodged in the county jail at Haddington, charged with a trespass in the churchyard of that town. So enraged was the mob against them, that an[Pg 28] attempt was made to force the jail in order to get at them. On their way to the court, the men were again attacked, forced from the carriage, and severely maimed. After examination they were admitted to bail; but, when set at liberty, they were assailed with more violence than ever, and were nearly killed. On the 29th of June, 1823, being Sunday, a most extraordinary outrage was perpetrated in the streets of Edinburgh. A coach containing an empty coffin and two men, was observed proceeding along the south bridge. The people suspecting that it was intended to convey a body taken from some churchyard, seized the coach. It was with difficulty that the police protected the men from the assaults of the populace: the coach they had no power to preserve. The horses were taken from it, and together with the coffin, after having been trundled a mile and a half through the streets of the city, it was deliberately projected over the steep side of the mound, and smashed into a thousand pieces. The people following it to the bottom, kindled a fire with its fragments, and surrounded it like the savages in Robinson Crusoe, till it was entirely consumed. In this case there was no foundation for their suspicions. The coffin was intended to have conveyed to his house in Edinburgh the body of a physician who that morning had died in a cottage near the neighborhood. A similar assault was some time ago made on two American gentlemen, who went to visit the abbey of Linlithgow after nightfall. The churchyards of the "gude Scots" are now strictly guarded by men and dogs; watch-towers are erected within the grounds, and *mort-safes*, as they are called, that is to say, strong iron frames are deposited in the ground over the graves. These people sometimes declare that they will put an end to anatomy, and certainly they are succeeding in the accomplishment of this menace as rapidly as they can well desire. The average number of medical students in Edinburgh is 700 each session. For several years past the difficulty of procuring subjects in that place has been so great, that out of all that number, not more than 150 or 200 have ever attempted to dissect; and even these have latterly been so opposed in their endeavours to prosecute their studies, that many of them have left the place in disgust. We have been informed by a friend, that he alone was personally acquainted with twenty individuals who retired from it at the beginning of last session, and who went to pursue their studies at Dublin, and we know that vast numbers followed their example at the end of the winter course. The medical school at Edinburgh, in fact, is now subsisting entirely on its past reputation;

in the course of a few years it will be entirely at an end, unless the system be changed. Let those who have the prosperity of the [Pg 29]university at heart, and who have the power to protect it, consider this before it be too late: they may be assured it is no idle prediction; for we give them notice, that it is at this moment the universal opinion and the current language of every well-informed medical man in England.

An excellent system of anatomical plates, which has been well received by the profession, has lately been published by Mr. Lizars, a lecturer on anatomy and physiology, in Edinburgh. This gentleman states that he has been induced to undertake this work, in order to obviate the most fatal consequences to the public; as far, at least, as a reference to art, instead of nature, is capable of obviating those consequences. He affirms, that the difficulty of obtaining instruction from nature has risen to such a pitch, owing to the extraordinary severity exercised by the legal authorities of the kingdom against persons employed in procuring subjects for dissection, as to threaten the ultimate destruction of medical and anatomical science. In his preface to the second part of his work, he apologizes to his readers for dividing one portion of it from another, with which it ought to have been connected; but states that he has been compelled to do so from the prejudices of the place, which prevented him for upwards of five months, from procuring a subject from which he might make his drawings. "In place of living," he says, "in a civilized and enlightened period, we appear as if we had been thrown back some centuries into the dark ages of ignorance, bigotry and superstition. Prejudices, worthy only of the multitude, have been conjured up and appealed to, in order to call forth popular indignation against those whose business it is to exhibit demonstratively the structure of the human body, and the functions of its different organs. The public journals, from a vicious propensity to pander to the vulgar appetite for excitement, have raked up and industriously circulated stories of exhumation of dead bodies, tending to exasperate and inflame the passions of the mob; and persons who, by their own showing, are friendly to the interests of science, have, in the excess of their zeal that bodies should remain undisturbed in their progress to decomposition, laboured to destroy in this country, that art, whose province it is to free living bodies from the consequences inseparable from accident and disease. And, which is worst of all, the prejudices of the multitude have been confirmed and rendered inveterate by the proceedings in our courts of justice, which have visited with the punishment due only to felons, the unhappy persons necessarily employed in the present state of the law, in procuring subjects for the dissecting-room."

He then goes on to state, that until anatomy be publicly[Pg 30] sanctioned in Edinburgh, the school of medicine there can never flourish; that upon the present system, young men obtain a degree or a diploma after a year or two of grinding, that is, of learning by rote the answers to the questions which the examiners are in the habit of putting to the candidates; that ignorant of the very elements of their profession, numbers of persons thus educated annually, go to the East and West Indies, and to the army and navy, where they have the charge of hundreds of their suffering fellow creatures, to whom they are in fact the instruments of cruelty and murder. In the preface to the 4th Part, he adds, that when Part II. was published, in the early part of the session, he took occasion to express his sorrow for the degraded state of his profession, and the threatened ruin of the Medical School of his native place, owing to the scarcity of subjects: That, for doing this, he has incurred considerable censure: that he regrets that he has yet found no reason to alter his opinion, for the winter session is now near its conclusion, and, he candidly declares, that such has been the scarcity of material, that *no teacher of anatomy or surgery has been able either to follow the regular plan of his course, or to do his duty to his pupils*; the consequence of which has been, that many of the students have left the school in disgust, and gone either to Dublin or Paris; while a still greater number, deprived of the means of dissecting, have contented themselves with lectures or theories, and with grinding; and entered on the practice of their profession ignorant of its fundamental principles.

Much of this opposition on the part of the people, arises from the present mode of procuring subjects. Fortunately, there is in Great Britain no custom, no superstition, no law, and we may add, no prejudice, against anatomy itself. There is even a general conviction of its necessity; there may be a feeling that it is a repulsive employment, but it is commonly acknowledged that it must not be neglected. The opposition which is made, is made not against anatomy, but against the practice of exhumation: and this is a practice which ought to be opposed. It is in the highest degree revolting; it would be disgraceful to a horde of savages; every feeling of the human heart rises up against it: so long as no other means of procuring bodies for dissection are provided, it must be tolerated; but, in itself, it is alike odious to the ignorant and the enlightened, to the most uncultivated and the most refined.

But the capital objection to this practice is, that it necessarily creates a crime, and educates a race of criminals.—Exhumation is forbidden by the law. It is, indeed, prohibited by no statute, either in England or Scotland: in both, it is an offence punishable at common law. There is a [Pg 31]statute of James the first, which makes it felony to steal a dead body for the purpose of witchcraft; there is none against taking a body for the purpose of dissection. In the case of the King against Lynn (1788), the court decided that the body being taken for the latter purpose, did not make it less an indictable offence; and that it is without doubt cognizable in a criminal court, because it is an act "highly indecent, at the bare idea of which nature revolts." It is punishable, therefore, by fine or imprisonment, or both: In Scotland, it is also punishable by whipping, and even by transportation.

We expected better things of America. We cannot express our astonishment and indignation, when we found that the state of New York has actually made it felony to remove a dead body from the place of sepulture for the purpose of dissection, without providing in any other mode for the schools of anatomy. This is worse than any thing that exists in any other part of the world. If these pages should meet the eye of any of our American brethren, we intreat them to read with attention, the facts which have been stated in the former part of this article, and to consider with seriousness the mischief they are doing. It will not be believed in England, that such scenes could have been witnessed in America, as were actually exhibited there scarcely a month ago. To satisfy our readers, however, that we do not misrepresent the state of things in that country, we transcribe the following accounts from *The New York Evening Post*, of *May 20th*. "At the late Court of Sessions, Solomon Parmeli was indicted for a misdemeanor, in entering Potter's Field, and removing the covers of two coffins deposited in a pit, and covered partly with earth. *The statute of this state making it a felony, to dig up or remove a dead human body with intent to dissect it*, did not embrace this case; because the prisoner had not dug up or removed the body. Mr. Schureman, the present keeper of Potter's Field, suspected that some person had entered it for the purpose of removing the dead; and, after sending for two watchmen, and calling his faithful dog, he went to ascertain the fact. On arriving at the grave, he found his suspicion confirmed; and requested the person concealed in the pit, to come out and show himself: no answer being given, Mr. Schureman sent his dog into the pit, and in the twinkling of an eye a tall stout fellow made his appearance, and took to his heels across the field. The night being dark, he might have effected his escape, had it not been for the sagacity and courage of the dog, who pursued him for some distance, but at last came up with him, seized and held him fast, until the arrival of Mr. Schureman and the watchmen, who secured him. The jury convicted the prisoner, and the[Pg 32] court sentenced him to six months' imprisonment in the Penitentiary. *The young gentlemen attending the Medical School of this city, will take warning by this man's fate. They may rest assured, that the keeper of Potter's Field will do his duty, and public justice will be executed on any man, whatever may be his condition in life, who is found violating the law, and the decency of Christian burial!*" The same paper gives the following account of a transaction, which took place at Hartford, in Connecticut, May 17. "Yesterday morning, two ladies were taking a walk in the South burying ground, when they discovered a tape-string, and a piece of cloth, which upon examination was found to be the piece that was laced upon Miss Jane Benton's face, who came to her death by drowning, and was buried a few days since. The ladies then went to the grave, and found that it had been disturbed—that she was taken out of her coffin, and a rope around her neck. The circumstance has produced great excitement in the public mind; and every one is on the alert to discover the perpetrators of this unfeeling, brutal act. *The citizens turned out in a body yesterday, and interred the corpse again!*"

These scenes are highly disgraceful, and disgraceful to all, though not *alike* to all parties. We do not blame the Americans for abolishing the practice of exhumation; but we blame them for stopping there. We maintain, that it is both absurd and criminal, to make this practice felony, without providing in some other method for the cultivation of anatomy.

In Great Britain, the law against the practice of exhumation is not allowed to slumber. There may be other cases which have not come to our knowledge; but we have ascertained that there have been 14 convictions for England alone, during the last year. The punishments inflicted have been imprisonment for various periods, with fines of different sums. The fines in general are heavy, considering the poverty of the offenders. Several persons are, at this moment, suffering these penalties; among others, there is now in the gaol of St. Alban's, a man who was sentenced for this offence to two years' imprisonment, and a fine of twenty pounds. The period of his confinement has expired some time; but he still remains in prison, on account of his inability to pay the fine.[1] Since the passing of the new Vagrant act, it has been the common practice to commit these offenders to hard labour for various periods. Very lately, two men, convicted of this offence, were sent to the Tread Mill, in Cold Bath Fields; one of whom died in one month[Pg 33] after his commitment. It is an error to suppose that these punishments operate to prevent exhumation; their only effect is to raise the price of subjects: a little reflection will show that they can have no other operation. At present, exhumation is the only method by which subjects for dissection can be procured; but subjects for this purpose must be procured: and be the difficulties what they may, will be procured: diseases will occur, operations must be performed, medical men must be educated, anatomy must be studied, dissections must go on. Unless some other means for affording a supply be adopted; whatever be the law or the popular feeling, neither magistrates, nor judges, nor juries, will, or can, put an entire stop to the practice. It is one, which, from the absolute necessity of the case, must be allowed. What is the consequence? So long as the practice of exhumation continues, a race of men must be trained up to violate the law. These men must go out in company for the purpose of nightly plunder, and plunder of the most odious kind, tending in a peculiar and most alarming measure to brutify the mind, and to eradicate every feeling and sentiment worthy of a man. This employment becomes a school in which men are trained for the commission of the most daring and inhuman crimes. Its operation is similar, but much worse than the nightly banding to violate the game laws, because there is something in the violation of the grave, which tends still more to degrade the character and to harden the heart. This offence is connived at; nay, it is rewarded; these men are absolutely paid to violate the law; and paid by men of reputation and influence in society. The transition is but too easy to the commission of other offences in the hope of similar connivance, if not of similar reward.

It is an odious thing that the teachers of anatomy should be brought into contact with such men: that they should be obliged to employ them, and that they should even be in their power; which they are to such a degree, that they are obliged to bear with the wantonness of their tyranny and insult. All the clamour against these men, all the punishment inflicted on them, only operate to raise the premium on the repetition of their offence. This premium the teachers of anatomy are obliged to pay, which these men perfectly understand, who do not at all dislike the opposition which is made to their vocation. It gives them no unreasonable pretext for exorbitancy in their demands. In general, they are men of infamous character; some of them are thieves, others are the companions and abettors of thieves. Almost all of them are extremely destitute. When apprehended for the offence in question, the teachers of anatomy are obliged to pay the expenses of the trial, and to support[Pg 34] their families while they are in prison: whence the idea of immunity is associated, in these men's minds, with the violation of the law, and when they do happen to incur its penalties, they practically find that they and their families are provided for, and this provision comes to them in the shape of a reward for the commission of their offence. The operation of such a system on the minds of the individuals themselves is exceedingly pernicious, and is not a little dangerous to the community.

Moreover, by the method of exhumation, the supply after all is scanty; it is never adequate to the wants of the schools; it is of necessity precarious, and it sometimes fails altogether for several months. But it is of the utmost importance that it should be abundant, regular, and cheap.—The number of young men who come annually to London for the purpose of studying medicine and surgery, may be about a thousand. Their expenses are necessarily very considerable while in town; they have already paid a large sum for their apprenticeship in the country; the circumstances of country practitioners, in general, can but ill afford protracted expenses for their sons in London; few of them stay a month longer than the time prescribed by the College of Surgeons. But the short period they spend in London, is the only time they have for acquiring the knowledge of their profession. If they mispend these precious hours, or if the means of employing them properly be denied them, they must necessarily remain ignorant for life. After they leave London they have no means of dissecting. We have seen that it is by dissecting alone, that they can make themselves acquainted even with the principles of their art; that without it they cannot so much as avail themselves of the opportunities of improvement, which experience itself may offer, nor, without the highest temerity, perform a single operation. We have seen that occasions suddenly occur, which require the prompt performance of important and difficult operations; we have seen that unless such operations are performed immediately, and with the utmost skill, life is inevitably lost. In many such cases, there is no time to send for other assistance. If a country practitioner (and most of these young men go to the country) be not himself capable of doing what is proper to be done, the death of the patient is certain. We put it to the reader to imagine what the feelings of an ingenuous young man must be, who is aware of what he ought to do, but who is conscious that his knowledge is not sufficient to authorise him to attempt to perform it, and who sees his patient die before him, when he knows that he might be saved, and that it would have been in his own power to save him, had he been properly educated. We put it to the reader to [Pg 35]conceive what his own sensations would be, were an ignorant surgeon, with a rashness more fatal than the criminal modesty of the former, to undertake an important operation—Suppose it were a tumor, which turned out to be an aneurism; suppose it were a hernia, in operating on which the epigastric artery were divided, or the intestine itself wounded: suppose it were his mother, his wife, his sister, his child, whom he thus saw perish before his eyes, what would the reader then think of the prejudice which withholds from the surgeon that information, without which the practice of his profession is murder?

The study of anatomy is a severe and laborious study; the practice of dissection is on many accounts highly repulsive: it is even not without danger to life itself.[2] To men of clear understandings, to those especially of a philosophical turn of mind, the pursuit is its own reward; they are so fully satisfied, that the more it is cultivated the more satisfaction it will afford, that they need no stimulus to induce them to undergo the drudgery. But this is by no means the case with ordinary minds. The fatigue and disgust of the dissecting-room, are appalling to them, and they need the stimulus of necessity to urge them to the task. The court of examiners of the College of Surgeons, requires from the candidates for surgical diplomas certificates that they have gone through at least two courses of dissections; the examiners at Apothecaries'-hall do not require such certificates. The consequence is, that many young men content themselves with attending lectures, and with passing their examinations at Apothecaries' hall, and do not apply for a diploma at the College of Surgeons. This single fact is sufficient to demonstrate to the public, that instead of throwing obstacles in the way of dissection, it is a duty which they owe to themselves to afford every possible facility to its practice, and to hold out to every member of the profession, the most powerful inducements to engage in it, by rewarding with confidence those who cultivate anatomy, by making excellence in anatomy indispensable to all offices in dispensaries and hospitals, and by thus rendering it impossible for any one who is ignorant of anatomy, to obtain rank in his profession. When a candidate presents himself for a diploma in Denmark, in his first trial he is put into a room with a subject, a case of instruments, and a memorandum, and informed that he is to display the anatomy of the face and neck, or that of the upper extremity or that of the lower extremity: that by the anatomy is to be understood, the blood-vessels, nerves, and muscles; and that as[Pg 36] soon as he has accomplished his task, the professors will attend his summons to judge of his attainments. These professors are the true examiners!

We shall have entered into the discussion of this subject to little purpose, if we have not produced in the minds of our readers a deep conviction, that anatomy ought to form an essential part of medical education, that anatomy cannot be studied without the practice of dissection; that dissection cannot be practised without a supply of subjects, and that the manner in which that supply is obtained in England is detestable, and ought immediately to be changed. It might be changed easily. We agree with Mr. Mackenzie, that legislative interference is necessary; we are satisfied that nothing will be done in England without it. The plan which Mr. Mackenzie suggests is as follows: 1. That the clause of our criminal code, by which the dissection of the dead body is made part of the punishment for murder, be repealed. 2. That the exhumation of dead bodies be punishable as felony. 3. That no diploma in medicine or surgery, be granted by any faculty, college, or university, except to those persons who shall produce undoubted evidence of their having carefully dissected at least five human bodies. 4. That in each of the hospitals, infirmaries, work-houses, poor-houses, foundling-houses, houses of correction, and prisons of London, Edinburgh, Glasgow, and Dublin, and if need be, of all other towns in Great Britain and Ireland, an apartment be appointed for the reception of the bodies of all persons dying in the said hospitals, infirmaries, work houses, poor-houses, foundling-houses, houses of correction, and prisons, *unclaimed by immediate relatives, or whose relatives decline to defray the expenses of interment.* 5. That the bodies of all persons dying in these towns, and, if need be, in all other towns, and also in country parishes, *unclaimable by immediate relatives, or whose relatives decline to defray the expenses of interment,* shall be conveyed to a mort-house appointed in the said towns for their reception. 6. That no dead bodies shall be delivered from any hospital, infirmary, work house, poor-house, foundling-house, house of correction, prison, or mort-house for anatomical purposes, except upon the requisition of a member of the Royal College of Physicians or of Surgeons, of London, Edinburgh, or Dublin, or of the Faculty of Physicians and Surgeons of Glasgow, and upon the payment of twenty shillings into the hands of the treasurer, of the hospital, infirmary, work-house, poor-house, foundling-house, house of correction, prison, or other officer appointed to receive the same. [This is too large a sum.] 7. That no dead body shall be conveyed from a hospital, infirmary, work-house, poor-house, foundling-house, house[Pg 37] of correction, prison, or mort-house, to a school of anatomy, except in a covered bier, and between the hours of four and six in the morning. 8. That after the expiration of twenty-eight days, an officer appointed for this purpose, in each of the four towns above-mentioned, shall cause the remains of the dead to be placed in a coffin, removed from the school of anatomy, where the dead body has been examined, to the mort-house of the town and decently buried. 9. That the expenses attending the execution of these regulations, be defrayed out of fees paid by teachers and students of anatomy, on receiving dead bodies from the hospitals, infirmaries, work-houses, poor houses, foundling-houses, houses of correction, prisons, and mort-houses.

To this plan there is but one objection, viz. that it is making the bodies of the poor public property. The answer is, that the limitation in the proposed law, which the objection does not notice, entirely removes the weight of that objection. Though no maxim can be more indisputable than that those who are supported by the public die in its debt, and that their remains at least, might, without injustice, be converted to the public use, yet it is not proposed to dispose in this manner of the bodies of all the poor: but only of that portion of the poor who die unclaimed and without friends, and whose appropriation to this public service could, therefore, afford pain to no one. If any concession and co-operation on the part of the public, for this great public object is to be expected, and without concession and co-operation nothing can be done, it is not easy to conceive of any plan which requires less public concession or implies less violation of public feeling. In point of fact it would put no indignity, it would inflict no injury on the poor; it is the rejection of it that would really and practically be unjust and cruel. The question is, whether the surgeon shall be allowed to gain knowledge by operating on the bodies of the dead, or driven to obtain it by practising on the bodies of the living. If the dead bodies of the poor are not appropriated to this use, their living bodies will and must be. The rich will always have it in their power to select, for the performance of an operation, the surgeon who has already signalized himself by success: but that surgeon, if he have not obtained the dexterity which ensures success, by dissecting and operating on the dead, must have acquired it by making experiments on the living bodies of the poor. There is no other means by which he can possibly have gained the necessary information. Every such surgeon who rises to eminence, must have risen to it through the suffering which he has inflicted, and the death which he has brought upon hundreds of the poor. The effect of the entire abolition of the practice of dissecting the dead, would be, to convert[Pg 38] poor-houses and public hospitals into so many schools where the surgeon, by practising on the poor, would learn to operate on the rich with safety and dexterity. This would be the certain and inevitable result: and this, indeed, would be to treat them with real indignity, and horrible injustice;

and proves, how possible it is to show an apparent consideration for the poor, and yet practically to treat them in the most injurious and cruel manner.

Nor would the proposed plan be the means of deterring this class of people from entering the hospitals. There is something reasonable in the apprehension on which this objection is founded: but the answer to it is complete, because it is an answer, derived from experience, to an objection, which is merely a deduction from what is probable. The plan has been acted on, and found to be unattended with this result: it was tried in Edinburgh, and the hospital was as full as it is at present: it is universally acted on in France, and the hospitals are always crowded.

The great advantages of the plan are, that it would accomplish the proposed object, easily and completely, whereas the plan in operation effects it imperfectly and with difficulty; and it would put an immediate and entire stop to all the evils of the present system. At once it would put an end to the needless education of daring and desperate violators of the law. It would tranquillize the public mind. Their dead would rest undisturbed: the sepulchre would be sacred: and all the horrors which the imagination connects with its violation would cease for ever.

We have stated, that the plan has been tried. Experience has proved its efficacy. It was adopted with perfect success in Edinburgh more than a century ago. In the Council Register for 1694, it is recorded that all unclaimed dead bodies in the charitable institutions or in the streets, were given for dissection to the College of Surgeons, to one or two of its individual members, and to the professor of anatomy. This regulation, at that period, excited no opposition on the part of the people, but effectually answered the desired object. All the medical schools on the continent are supplied with subjects, by public authority, in a similar manner. We have obtained from a friend in Paris, a gentleman who is at the head of the anatomical department in that city, the following account of the manner in which the schools of anatomy are supplied. It is stated; 1. That the faculty of medicine at Paris is authorized to take from the civil hospitals, from the prisons, and from the depôts of mendicity, the bodies which are necessary for teaching anatomy. 2. That a gratuity of eight pence is given to the attendants in the hospitals for each body. 3. That upon[Pg 39] the foundation by the National Convention, of schools of health, the statutes of their foundation declare, that the subjects necessary for the schools of anatomy shall be taken from the hospitals, and that since this period, the council of hospitals and the prefect of police, have always permitted the practice. 4. That M. Breschet, chief of the anatomical department of the faculty of Paris, sends a carriage daily to the different hospitals, which brings back the necessary number of bodies: that this number has sometimes amounted to 2000 per annum for the faculty only, without reckoning those used in L'Hôpital de la Pitié, but that since the general attention which has recently been bestowed upon pathologic anatomy, numbers of bodies are opened in the civil and military hospitals, and that the faculty seldom obtain more than 1000 or 1200. 5. That, besides the dissections by the faculty of medicine, and those pursued in L'Hôpital de la Pitié, theatres of anatomy are opened in all the great hospitals, for the pupils of those establishments: that in these institutions anatomy is carefully taught, and that pupils have all the facilities for dissection that can be desired. 6. That the price of a body varies from four shillings to eight shillings and sixpence. 7. That after dissection, the bodies are wrapt in cloths, and carried to the neighbouring cemetery, where they are received for ten-pence. 8. That the practice of exhumation is abolished: that there are insurmountable obstacles to the return of that system, and that bodies are never taken from burial grounds, without an order for exhumation, which is given only when the tribunals require it for the purpose of medico-legal investigation. 9. That though the people have an aversion to the operations of dissection, yet they never make any opposition to them, provided respect be paid to the laws of decency and salubrity, on account of the deep conviction that prevails of their utility, 10. That the relatives of the deceased seldom or never oppose the opening of any body, if the physicians desire it. That all the medical students in France, with scarcely any exception, dissect, and that that physician or surgeon who is not acquainted with anatomy, is universally regarded as the most ignorant of men.

It is time that the physicians and surgeons of England, should exert themselves to change a system which has so long retarded the progress of their science, and been productive of so much evil to the community. We are persuaded, that there is good sense enough, both in the people and in the legislature, to listen to their representations. We would advise them to avail themselves of the means they possess to communicate information to the people, and to make individual members of parliament acquainted with the subject. With this view we would recommend the whole[Pg 40] body to act in concert, to appoint a committee for conducting the matter, and to petition parliament, as soon as they shall have made the nature of their claims, and the grounds on which they rest, more generally known. If they act in co-operation with each other, and pursue their object temperately, and steadily, we cannot but believe, that their efforts at no distant period, will be crowned with success.

FOOTNOTES:

[1] Since the above was written, we have learned that this man has been recently liberated, and his fine remitted.

[2] A winter never passes without proving fatal to several students who die from injuries received in dissection.

Milton Keynes UK
Ingram Content Group UK Ltd.
UKHW040816051024
449151UK00004B/256